Human Centered Design in Success of Organization

Illustrated examples from surveyed studies

SUMMARY

This book presents a survey of studies where ergonomic intervention has improved the human condition and quality of life, especially at work with regard to safety and health outcomes, discomfort and enjoyment. These studies show proven economic advantages in terms of reduction of effort and cost expended

Implementation of human centered designs have been proven to Increase productivity, quality, management utilization and indeed decrease in number of lost work days, discomfort, recordable rates, fatigue and stress in workplaces.

The OBJECTIVE is to identify, determine and show the benefits of human centered design interventions in successes of organizations.

INTRODUCTION

This survey identified that ergonomics is a human centered design. It is the process of designing

and/or modifying tools, materials, equipment, work spaces, tasks, jobs, products, systems and

environment to match the mental and physical abilities, limitations and social needs of all people

affected. The purpose for starting and maintaining effective human centered design interventions

in workplaces is to address poor ergonomic quality from viewpoints including injury, discomfort,

quality, usability, productivity and profitability of products, process and services. For effective

intervention there should be management support, and people in the organization must be aware

and support the goals and scope of the intervention.

In this review ergonomists and his change team are charged with finding the many opportunities

for improving the working environment in relation to musculoskeletal system diseases. This

includes general problem identification, assessment and solution generation and evaluation using

tools and techniques from the field of ergonomics. There are many ways in which ergonomic

problems can be identified. These can range from general observations and checklists to quantitative risk assessment tools. Ideally, several approaches should be used: talking to employees and seeking their views. Employees have important knowledge of the work they do, any problems they have, and their impact on health, safety, and performance. Assessing the work system by asking questions such as: Is the person in a comfortable position? Does the person experience discomfort, including aches, pain, fatigue, or stress? Is the equipment appropriate, easy to use and well maintained? Is the person satisfied with their working arrangements? Are there frequent errors? Are there signs of poor or inadequate equipment design, such as plasters on workers' fingers or 'home-made' protective pads made of tissue or foam? Examining the circumstances surrounding frequent errors and incidents where mistakes have occurred and people have been injured. Use accident reports to identify details of incidents and their possible causes; recording and looking at sickness absence and staff turnover levels. High numbers may result from the problems listed earlier and/or dissatisfaction at work.

Once ergonomic problems have been identified a minor alteration may be all that is necessary to make a task easier and safer to perform. For example: provision of height-adjustable chairs so individual operators can work at their preferred work height; removal of obstacles from under desks to create sufficient leg room; arrangement of items stored on shelving so those used most frequently and those that are the heaviest are between waist and shoulder height; raising platforms to help operators reach badly located controls; changing shift work patterns; and introduction of job rotation between different tasks to reduce physical and mental fatigue.

Talking to employees and get them to suggest ideas and discuss possible solutions. Involving

employees from the start of the process - this will help all parties to accept any proposed

changes. Always it is good to make sure that any alterations are properly evaluated by the people

who do the job, and therefore care must be taken to ensure that a change introduced to solve one

problem does not create difficulties elsewhere.

EXAMPLES OF THE IMPACT OF HUMAN CENTERED DESIGN

1.0 Productivity

Several case studies have shown the positive implications of human centered design/ergonomic intervention for labor productivity. Many of these design efforts were aimed at reducing effects of mechanical exposure to operators and therefore improve their overall health associated with Musculoskeletal Disorders/Diseases (MSDs). Three such cases have been identified which were conducted in different production workstations and they include: -

(a) Amina Hameed et al (2009) that investigated the Impact of Office Design on Employees' Productivity at Banking Organizations.

(b) Debabneh et al, (2001) investigated the impact of frequent short rest breaks on the productivity and wellbeing of workers and

(c) Dempsey et al. (2002) investigated the effect of two different plier designs on human performance and discomfort. Their methods and findings are illustrated in table 1 below.

Table 1: How interventions improved labor productivity

Researcher	Procedure	Findings
Amina Hameed et al 2009)	Investigated the Impact of Office Design on Employees' Productivity at Banking Organizations. Data was collected through questionnaire. Observation was also used to collect information about the office design. The Questionnaire consisted of 24 questions; 4 questions on each variable. Out of these, 4 questions were on productivity, based on the technique of subjective productivity measurement. Subjective productivity data was gathered from the employees, supervisors, clients, customers and suppliers. A direct subjective productivity measurement is a survey question concerning an employees' own productivity. A 5-point Likert Scale was used to measure all the variables. The scale varies from 1 (strongly disagree) to 5 (strongly agree) for most of the questions. A few questions were measured by the 5-point Likert Scale ranging from 1 (not at	It was shown that office design has a substantial impact on the employees' productivity. The overall impact of different elements showed that lighting affects the productivity of most employees. There was a direct relationship between office design and productivity. The Relationship between Office design and Productivity was determined by using the Pearson's Correlation in SPSS. A strong correlation exists between elements of office design and productivity of office design. The regression analysis of the data shows that the coefficient of determination R. square = 0.576, so, it can be concluded that 58 percent of the variability in employees'

	all) to 5 (always). The questions in the questionnaire for the subjective productivity measurement were in percentages.	productivity is accounted for by the variables in this model.
Dababneh et al. (2001)	Investigated the impact of frequent short rest breaks on the productivity and wellbeing of a group of 30 workers in a meat-processing plant. Two rest break schedules were tested, both of which provided 36 min of extra break time over the regular break schedule. In the first experimental workers were given 12 3-min breaks evenly distributed over the workday. In the second schedule, workers were given four 9-min breaks evenly distributed over the workday. Outcome measures included production rate and discomfort and stress ratings. To measure discomfort operators were asked to complete a detailed body map with a 5 point scale which ranged from 0 (no discomfort) to 5 (extreme discomfort).	While there was a decrease in discomfort for both rest break conditions, the decrease in discomfort for the 9 minute break was significant when compared with no rest breaks. Discomfort decreased by approximately 17% for the 9 minute combination. Results concluded that the addition of rest breaks improved productivity by 25% to 30 %.
Dempsey et al. (2002)	Investigated the effect of two different plier designs on human performance and discomfort. There were 90 experimental	The experimenters found that higher work heights along with higher work-piece orientations increased

| | combinations, two plier types, 5 work heights and 9 work piece orientations in the sagittal plane. Treatments consisted of 5 one minute replications, 5 minute rest period and a 10 second pause between replications. Individuals stood in front of a work piece carriage which could be adjusted for different combinations to perform the task which consisted of turning a screw that was fixed in the rig. Individuals were asked to perform the task as quickly as possible to maximize the number of deviations in the allotted time. Productivity was measured as the number of revolutions the bolt was turned. Discomfort was rated using body map 7-point scale based on Corlett and Bishops (1976) | elbow heights leading to higher levels of discomfort in the shoulder and upper arm and this in turn led to lower productivity for these combinations. For example a discomfort score of approximately 2.5 for elbow height + 25 cm with +60Â° orientation in the sagittal plane; productivity was 16 revolutions per minute. Where the elbow height was not elevated and there was no shoulder orientation in the sagittal plane discomfort was found to be almost zero; productivity was just over 25 revolutions per minute. |

The illustration from table 1 show the positive benefits to the respective organization(s) derived from the interventions; not only did discomfort reduce in the operators, but it led to increase in productivity. In other words these studies did show that an increase in operator productivity and efficiency is related to the health of an operator. Productivity may be measured by objective

means such as number of exertions per minute (Observed Work Productivity) or by subjective

means (Perceived Work Productivity) where the individual is asked to rate the effect they believe

discomfort is having on their productivity through collecting data by use of questionnaires or

surveys. Data in these studies is based on Likert scale type questions, probably the most widely

used response scale featured in surveys might be a 5-point Likert scale ranging from "no

discomfort" (0) to "extreme discomfort" (5); or on 7-point Likert scale with more responses

added into the scale as shown in the case studies in table 3. Questionnaires or surveys are used

because they determine an indirect measure of productivity by giving an overall productivity

score to each participant.

The actual effects of productivity in relation to operator health are expressed in terms of

absenteeism and presenteeism. Data on absenteeism may be obtained from the human resource

records prior to the distribution of the questionnaires. Only involuntary absences (i.e., illness and

injury) should be included in the study. Absenteeism is measured by the number of hours of

involuntary absences recorded from the period being investigated.

Absenteeism may be defined as time missed from work due to health problems (Boles et al.,

2004), on the other hand Presenteeism may be defined through the costs associated with

productivity loss when the operator turns up to work, but does not engage in work as

productively as their peers due to distractions related to health issues or social pressures. The fact that the unhealthy operator cannot work as efficiently as their healthy counterpart implies that

there is a reduced labor productivity as this operator is unable to work at a normal rate and reach

maximum output. As a result overall industrial productivity is reduced.

Earlier researches focused on absenteeism as the actual cost of productivity losses, as lost days

are easily quantifiable, however research by Brouwer et al., 1999 estimated that 0.93% of all

working hours are lost due to on-site presenteeism. Stewart et al. (2003) estimated that

presenteeism accounted for 71% of the $226 billion worth of lost productive time per year. Both

absenteeism and presenteeism are therefore considered an integral part of productivity and

should be considered in tandem (Escorpizo, 2008).

2.0 Quality

Quality issues have been associated with poorly designed workstations and poor work

organization. Many authors have done research on this issue where better work design led to

better quality of products. This survey identified three cases studies;

(a) Eklud (1995) investigated the relationship between ergonomics and quality in assembly

work in a car manufacturing plant;

(b) Helander et al, (1995) performed an ergonomic redesign of a workstation in IBM which

identified several ergonomics problems and

(c) Lin et al. (2001) examined the relationship between workstation ergonomics and product

quality in a disposable camera factory. The methods used and findings are presented in

the table 2 below.

Table 2: How the interventions improved quality

Case study	Procedures	Findings
Eklund (1995)	Did a case study on a Swedish car manufacturing plant to investigate the relationship between ergonomics and quality in assembly work. Ergonomic data were obtained from work place analysis, interviews and questionnaires. Quality data were obtained from the company's data base and employee interviews. It was found that quality deficiencies were three times more prevalent in workstations with ergonomics issues. For example, there were ergonomics problems in 25% of the workstations investigated and these workstations accounted for 49% of all end product quality deficiencies. On the other hand, the remaining 75% of tasks made up only 51% of quality deficiencies.	Direct causes of quality deficiencies were found to be discomfort as a result of strained body parts as well as organizational issues such as time and pressure.

Helander et al (1995)	Performed an ergonomic redesign of a workstation in IBM which identified several ergonomics problems, including disorganized workstations, the operators were unable to sit, and the tasks composed of awkward postures for both the right and left hands. There were also insufficient illumination of the task and other arrangement issues.	After workstation redesign all of these problems were eliminated, worker comfort improved and there were contiguous improvements in *production quality*.
Lin et al. (2001)	Examined the relationship between workstation ergonomics and product quality, at two manual assembly lines in a disposable camera production factory. Line A was older and non-automated, whereas line B was newer and semi-automated. Quality on the assembly lines was measured as the number of defects per week. Postural data were also obtained from video tape analysis and analyzed as per	It showed the benefits of an ergonomic intervention in the area of quality and hence in any ergonomic design the operator should be an integral part of system design. Drury, (2000) says that differences have been noticed at the level of action at which each of the elements is implemented; quality is implemented at a strategic level while ergonomics is implemented at a technical level

	Drury (1987). The mean value of errors per week for line A was 23.4 with time take equal to 4.82 seconds and a posture score of 9. On the other hand, for lines B there were 4.02 errors per week, 3.84 seconds taken for the task and a posture score of 9.7 over a week using Drury's method. The use of Regression equations for the two lines was capable of predicting over 50% of the quality variance which were highly significant.	

Illustrated examples in table 2 show how good ergonomic designed work environment led to overall better quality. Increased quality means less defective products or rejected products and thus less wastage of both material and time of producing them. Increased quality means that the end product is appealing and satisfying to end users or customers, and so is the image of the organization. And another fact- it is costly for organizations to pay for producing products which cannot be sold; it is even more costly when raw material and time is wasted this way.

3.0 Cost

Several case studies have demonstrated the cost-benefits of implementing human centered

design. Three cases were identified demonstrating this:-

(a) Judy Village et al (1998) investigated cost-benefit analysis of an ergonomic intervention

in two hospital laundries versus a control laundry;

(b) Sen et al (2003) investigated the costs and benefits of ergonomic redesign of an

electronic motherboard in a company and

(c) Hendrick (1996) Investigated health and safety problems and associated costs at AT&T

Global Information solutions. Table 4 illustrates these interventions.

Table 3: Cost-benefits of human centered design/ergonomic interventions.

Author	Intervention	Cost savings
Judy Village et al, (1998) At Hospital Laundry	An ergonomic intervention study was conducted at two hospital laundries and a control laundry to evaluate changes in injuries, self-reported pain and psychosocial factors post-intervention. At the two	The total cost of ergonomic interventions at Laundry A was $84,240. Benefits totaled $82,070, including a reduction in managers' time dealing with grievances, selling of a sheet ironer, reduced turnover and

	intervention laundries A and B, detailed ergonomic assessments and measurements were made of each job. The post-intervention questionnaire also asked workers about the ergonomics process, such as whether their job had changed, whether it improved their job, reduced fatigue, etc. Questionnaire data was analyzed pre and post-intervention using either a chi square analysis or analysis of variance and Mann Witney tests. Follow-up interviews were conducted to collect outcome benefits such as changes in productivity, quality, injuries, overtime, ease of return-to-work and costs.	overtime and increased productivity. The one year benefit-to-cost ratio was 0.97. Therefore, the ergonomic solutions had a one-year payback or 97% return on investment. At Laundry B, the cost of ergonomic interventions was $27,800. Benefits included reduced MSIs, reduced time loss during return-to-work and an increase in productivity, totaling $41,600. The one-year benefit-to-cost ratio is 1.5, with an eight month payback for the ergonomic solutions. The return on investment for Laundry B is therefore 150%.
Sen et al (2003) Electronic motherboard company.	Investigated the costs and benefits of ergonomic redesign of an electronic motherboard in a company. Before the intervention the company was running at a loss due to high reject costs along with poor quality and	There was a direct savings of $ 581,495 per year and operator occupational health and safety improved.

	productivity. The analysts made ergonomic assessments and issues were identified as being associated with design problems of motherboards. A particular machine had problems in placing integrated circuits onto pads. The operators had issues manually soldering certain components which resulted in health and safety and cost issues. The boards were ergonomically redesigned to allow more of them to be machine soldered.	
Hendrick (1996) AT&T Global Solutions	Investigated health and safety problems and associated costs at AT&T Global Information solutions in San Diego with sample size of 800 people. Three types of frequent injury were identified from the manufacturing process; lifting, fastening and keyboarding. Work analysis was used to identify ergonomics problems and introduce solutions.	Extensive ergonomic workstation improvements were completed and compensation dropped from $400,000 to $94,000 per year. Subsequent interventions realized compensation losses dropping by a further $12,000 per year. Within four years lost days due to injury had dropped from 298 to zero and there was a saving of $1.48 million.

The three example in table 3 are among the many that have shown why ergonomic interventions are vital to savings of costs in organization if and when they are implemented. Organizations should be aware that while the costs associated with Musculoskeletal Disorders are considerable, the cost savings gained through implementing interventions are huge and hence they should not hesitate to implement them should the need arise. Ergonomists too shall ensure that they prepare a sound proposal that shows both the technical and cost sides of the benefits when presented to the organization for approval. As the benefits of ergonomic interventions are generally over and above initial cost savings, it is felt that general economic and accounting tools are unable to capture the true benefits of ergonomic interventions. Various cost-benefit ergonomic tools have been developed to highlight the benefits of ergonomic interventions.

Costs associated with Musculoskeletal Disorders are alarming, however direct costs represent only a fraction of total costs. It is estimated that total costs may be two to three times the direct compensation costs (Oxenburgh et al., 2004). Direct costs are mainly associated with health care and which include outpatient costs which can be further broken down into physician visits, medication and outpatient surgery. In-patient costs include acute hospital facilities use, with or without surgery. There are personal costs borne by the patient such as transport and care time. Other disease related costs include environmental adaptations and medical equipment to aid life-style changes for chronic illness effects.

For an organization indirect costs are generally associated with quality and Productivity. Lower quality products have been associated with Musculoskeletal Disorders pain; which in turn leads to increased product re-work and rejects which is a financial burden for an organization. Productivity related costs result from worker performance issues, for example an operator is not as physically capable of doing the work due to Musculoskeletal Disorders pain. An operator may also lose motivation if suffering from Musculoskeletal Disorders related pain.

An operator may be a less experienced one replacing another operator with MSD pain; so they will not perform as well as the more experienced worker would. On some cases more experienced workers may have to take time out to help a less experienced worker.

Cost is often a motivating factor for most organizations to prioritize projects. Ergonomic interventions are often viewed as industrial projects and hence compete with other projects for limited resources. Accounting calculations or methods are generally used to highlight potential cost savings. However, ergonomics projects have additional benefits over and above initial and direct cost savings which may not be accounted for in traditional cost tools. Moreover, since as ergonomist is often not a financial expert and he therefore may be limited in projecting cost benefit. Cost benefit models have been developed for this purpose to assess the benefits of ergonomics interventions. [Note: Another book will dedicate on Examples and Use of these cost benefit models]

SUMMARY and CONCLUSION

This book has identified successes of implementing human centered design in organization(s). The case studies surveyed are a preface of many researches that that have shown a direct link of the ergonomic interventions/ human centered design to the successes either through discomfort and cost reductions measures, injury and incidence rates management, quality and productivity improvements, and the overall cost, economic and performance improvements.

It can also be deduced from this book that Physical risk factors and their combinations are linked to the causation of Musculoskeletal Disorders (MSDs) which has implications on the performance improvements such as increased Productivity, comfort levels, Quality and decreased in Injury Cases, Fatigue & Stress, Lost work Days, and Recordable Rates in organizations.

REFERENCES

1) Journal of public affairs, administration and management Volume 3, Issue 1, 2009: Impact of Office Design on Employees' Productivity: A Case study of Banking Organizations of Abbottabad, Pakistan Amina Hameed, Shehla Amjad, 2009.

2) Research 40(16): 4059-4075. NIOSH (National Institute for Occupational Safety and Health) (1997). A Critical Review of Epidemiologic Evidence for Work-Related Musculoskeletal

3) ACGIH (2001). Documentation of the threshold limit values for physical agents. Ergonomics. In: Proceedings of the American Conference of Governmental Industrial Hygienists, Cincinnati, OH, USA. 7: 1-18.

4) Brouwer, W. B., Koopmanschap, M. A. and Rutten, F. F. (1999). Productivity losses without absence: measurement validation and empirical evidence. Health Policy 48: 13-27.

5) Christmansson, M. (1994). The HAMA method: a new method for the analysis of upper limb movements and risk for work-related musculoskeletal disorders. 12th Triennial Congress of the International Ergonomics Association/Human Factors Association of Canada Toronto, Human Factors Association of Canada: 173-175.

6) Dababneh, A. J., Swanson, N. and R.L., S. (2001). Impact of added rest breaks on the productivity and wellbeing of workers. Ergonomics 44 164-174.

7) Colombini, D. and Occhipinti, E. (2006). Preventing upper limb work-related musculoskeletal disorders (UL-WMSDS): New approaches in job (re)design and current trends in standardization. Applied Ergonomics 37: 441-450.

8) HandPak. http://www.wipergo.com/HandPakSoftware.htm.

9) Electronics manufacturing. International Journal of Industrial Ergonomics 15(1): 137-155.

10) Hendrick, H. W. (2003). Determining the cost-benefits of ergonomics projects and the factors that lead to their success. Applied Ergonomics 34: 419-427.

11) Presenteeism Scale: Health Status and Employee Productivity. Occupational and Environmental Medicine 44: 14-20.

12) Kumar, S. (2001). Theories of musculoskeletal injury causation. Ergonomics 44(1): 17 - 47.

13) Disorders of the Neck, Upper Extremity, and Low Back. Musculoskeletal

14) Disorders and Workplace Factors B. P. Bernard, NIOSH, National Institute for Occupational Safety and Health.

15) Occhipinti, E. (1998). OCRA: a concise index for the assessment of exposure to repetitive movements of the upper limbs. Ergonomics 41(9): 1290 - 1311.

16) Oxenburgh, M., Marlow, P., Oxenburgh, A., (2004). Increasing Productivity and Profit through Health and Safety. New York, CRC Press

17) Pousette, A. and Hanse, J. J. (2002). Job characteristics as predictors of ill-health and sickness absenteeism in different occupational types - a multigroup structural equation modelling approach. Work and Stress 16: 229-250.

18) Salvendy, G. (1992). Handbook of Industrial Engineering. New York, Wiley-Interscience in cooperation with: Institute of Industrial Engineers.

19) Sen., N. S., Yeow, P.H.P., (2003). Cost effectiveness of ergonomic redesign of electronic motherboards. Applied Ergonomics (34): 453-463.

20) Studies: New Procedures and Recommendations. Psychological Methods 7: 422-445.

21) Shrout, P. E. and Fleiss, J. L. (1979). Interclass correlations: uses in assessing rater reliability.

Psychological Bulletin 86(2): 420-428.

22) Disorders and Workplace Factors B. P. Bernard, NIOSH, National Institute for Occupational

Safety and Health

23) Kumar, S. (2001). Theories of musculoskeletal injury causation. Ergonomics 44(1): 17 - 47.

24) Kaplan, R. S. and Norton, D. P. (1992). The Balanced Scorecard: Measures that drive

performance. Harvard Business Review 70: 71-79.

25) HandPak. http://www.wipergo.com/HandPakSoftware.htm.

26) Goggins, R. W., Spielholz, P. and Nothstein, G. L. (2008). Estimating the effectiveness of

ergonomics interventions through case studies: Implications for predictive cost-benefit analysis.

Journal of Safety Research 39(3): 339-344.

27) Majorkumar Govindaraju et al, (1998) Quality improvement in manufacturing through human

performance enhancement. Integrated Manufacturing Systems 12/5 [2001] 360±367 # MCB

University Press[ISSN 0957-6061]